MW01073649

# JUSTIN THE HERO!

*May I introduce you to Justin? Justin is a hero. He saved a whole project! And by doing very simple things. But let´s start at the beginning. . . .*

# PROBLEMS

*Justin works as a project manager. A year ago he was managing a major software project. His team was facing some serious problems: Their plan was in danger of failing because they couldn't meet their deadlines. Justin tried forcing people to work overtime, and he even offered them bonuses so they would work more efficiently. These had amazingly little effect!*

# INSPIRATION

*One day Justin was printing a new project plan because the old one wasn't really useful anymore. To be honest, he did this quite often. As he looked at the printer, inspiration hit him! The printer finished one sheet of paper after another and loaded new sheets continuously. This was a nice, smooth process—new sheets pulled in on one side and completed printouts delivered on the other. "This looks like flow," he thought. Of course Justin wanted his printouts more quickly, but it was totally clear to him that the printer had only a limited printing capacity, and that it would be absolutely futile to cram more and more paper into it. Would it print faster? No! Such pressure would certainly lead to the opposite—poor quality—and sooner or later there would be a paper jam!*

# LIMITED CAPABILITY

*Nobody would ever try this with a printer. But it was exactly what Justin did in his project every single day: He crammed new tasks onto his team and urged them to go faster. But isn't it obvious that not only printers have limited capacity, but development teams as well? Generally speaking, one could say that every kind of knowledge work—software development, marketing, running a law firm or a publishing house, and so on—could be seen as a system with limited capacity, or rather, capability. If we push too much work into the system, it will overload, which leads to less efficiency and poor quality. Now it was totally clear to Justin what he needed to do: Identify the system's capability and make sure that it would never be exceeded!*

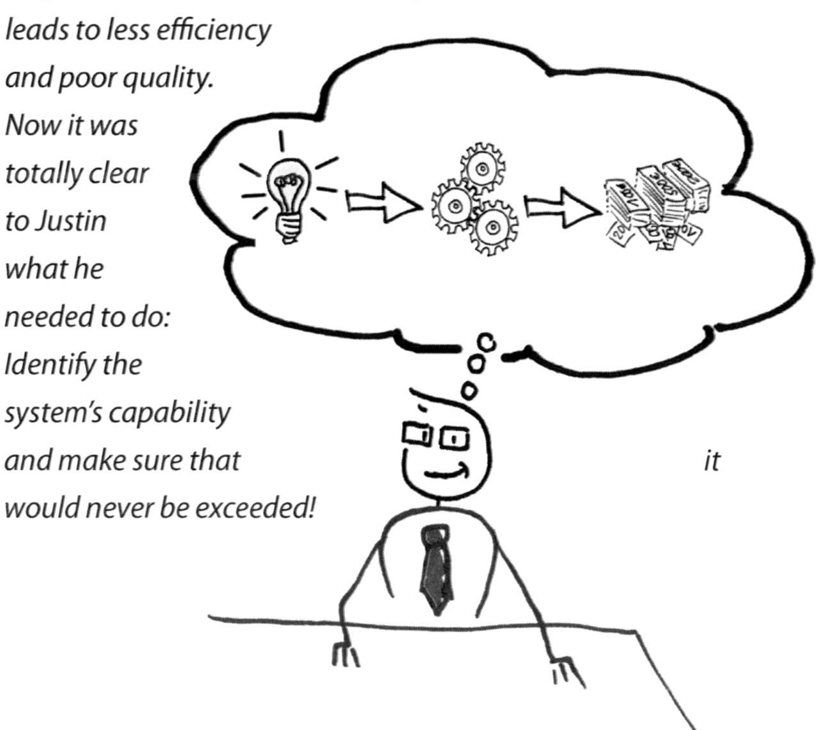

# VISUALIZATION

*But how could you identify the capability of a system for knowledge work? In manufacturing this is easy: If you're working above your capability, lots of stuff would pile up, which is easy to detect. In knowledge work, however, we're dealing with mostly abstract tasks. Usually we cannot see them, and therefore it's hard to detect overload. So the first step had to be to visualize the workflow and the tasks Justin's team needed to accomplish. Justin knew how this could be done because he'd already seen this in other teams in his organization.*

*He ordered a huge whiteboard and attached it to the wall of the team room. Every process step they were using (such as analysis, development, or testing) was visualized as a separate column. Every feature the team was working on became a sticky note that was placed in its respective column on the board. They put all the features that weren't started yet in the left-most column. They labeled that column "To Do." The column on the far right was called "Done." All completed tickets ended up there.*

Do
Analyze
Dev.
Test
Done

# OVERLOAD

*Justin's team knew that they had plenty of work to do, but what they saw now was frightening! The board was completely overloaded with tickets!*

*The good thing was that now they had a realistic picture of the state of their project—on a Kanban board. This seemed much better than the wish list they used to call a Gantt Chart!*

# WIP LIMITS

*Then Justin asked the analysts, developers, and testers, "How many tasks can you handle at the same time?" The analysts said three, the developers five, the testers two. These numbers were written over the column heads. In addition, they defined a simple rule: At no time should there be more tickets in a column than the number at the top of it. By doing this, they were limiting their WIP, which stands for Work in Progress.*

*They agreed on an overall WIP limit of ten for their whole system, meaning that they couldn't work on more than ten tasks at one time. Did this mean their capability was exactly ten? Definitely not, because they had simply guessed at these numbers. Still, their new system was much better than everything else they had come up with before. Then they decided to come together on a regular basis to reflect on the WIP limits and adjust them if necessary.*

# REDUCING WORK

*But now a new problem emerged: They had many more tickets in progress than their WIP limits allowed. The team discussed two options for how to solve this problem: The "hard way" would have been to stop tasks and reassign them to the "To Do" column. But they decided to take another path, which seemed to be smoother, and hence more appropriate for their situation. They kept all started tasks in the system—knowing they were exceeding their capability.*

*A policy was defined that said, "We don't start working on new tasks before we honor our limits." This policy was written on a flipchart and attached to the wall—right beside their Kanban board.*

# STOP STARTING

*This policy had quite a big impact on the team: They focused on finishing existing tasks and didn't start new work as often as they had been before. To coordinate their work, they met in front of the board every day. During these standup meetings, they walked the board from right to left, focusing on one card after another and asking: "What can we do to finish this ticket?" On the Internet, Justin had once read the Kanban slogan: "Stop starting, start finishing!"*

*This slogan was written above the board in big letters.*

# GETTING STUFF DONE

*Gradually the team managed to finish more and more tickets and so the board became emptier and emptier. After a couple of days they had finished enough tasks to clear the overflow, and all limits were respected. It was clear that they profited from this new way of working. The most obvious advantage was that they finished tasks much more quickly than they had before. At first Justin thought the only reason for this was that they had reduced multitasking. There was less task-switching, and hence less mental overhead for wrapping their heads around the same tasks over and over again.*

# LITTLE'S LAW

*But there was another very interesting thing Justin came across: There is this formula, called Little's Law. According to Little's Law, the average lead time for finishing tasks is calculated as the ratio between average Work in Progress and average Throughput. Work in Progress is work that is started but not yet finished. If one wants to shorten the lead time, there are two options: One way is to increase Throughput. This was exactly what Justin had tried to accomplish over and over again during the past few years—by buying better tools, forcing overtime, and yelling at employees—all to little avail! Reducing the Work in Progress, on the other hand, was quite simple and it had a huge impact!*

# QUALITY

*The second effect that Justin observed was improved quality as a result of shorter feedback cycles and less multitasking. In the book* Brain Rules *by John Medina, Justin had read that when processing several tasks at the same time not only do people need 50% more time to complete a task, but even worse—they make up to 50% more mistakes!*

# PULL SYSTEM

*The third advantage to setting WIP limits was that the team members experienced less overload than they had before and they enjoyed working again! Now there was an explicit limit for the work they were supposed to do – and nobody was allowed to push more and more tasks into the system.*

*The biggest challenge for Justin was that he had to hold back and trust that his team were doing their best. After a while there was no more doubt: They pulled in new work over and over again, and they were working better than ever before!*

# PREDICTABILITY

*Later on, Justin discovered another advantage: predictability. In their WIP-limited system, the amount of work stayed more or less constant all the time. It was much easier to predict when a single ticket would be finished, which helped him to establish Service Level Agreements for future projects with his stakeholders, as well as lightweight but useful release planning. Of course Justin knew that there is no such thing as 100% predictability— at least not in knowledge work.*

# FEEDBACK

*There was even a further advantage—one that became more and more important to Justin over time. With shorter lead times, the team received faster feedback for their work. Now they recognized sooner whether they had developed a system their customers were willing to pay for, if they had made the right assumptions, and where they had been mistaken. In a nutshell, now they could learn more quickly and more often, and they were able to use this information for their ongoing work.*

# UTILIZATION

*But something had to be wrong here! Sure, everything was going quite well, but the utilization of his staff went down dramatically! Because of the WIP limits, the work was stalling. It started at a certain point and eventually affected the entire system. It seemed that sooner or later team members wouldn't be allowed to continue working because of the WIP limits! This was a disaster. Like everybody else, Justin had learned that utilization is the most important criterion for optimization! Wasn't it?*

# SLACK

*Justin thought this through. The lead times for their tasks had been shortened, and the throughput of their system was higher than ever before. So did it really matter whether people were fully utilized? Actually, he liked this! The system was working better than before, and the team wasn't overloaded anymore. So why should this bother him? "It's better that people wait for work than the other way around," he thought. But it was even better than that! The team members used their slack time to help their colleagues and to think about general system improvements. One result of this was the development of the first automated acceptance tests. Now Justin understood what David Anderson meant by stating, "Slack is the ultimate weapon!"*

# VETERANS

*All continued going well, but after a while it became clear that the situation hadn't changed significantly for some team members. For some, the workload hadn't really gone down. There were two guys who had worked for this company for ages and who knew the system very well. Other team members approached them all the time to ask about the code, the architecture, and other things. Justin scheduled a workshop in order to have a closer look at the problem. The outcome was that the team decided to introduce another kind of limitation: personal limits. Little avatars were made to represent the two "project veterans," which were then attached to the tickets they were working on. Because there were only three avatars per person, the veterans couldn't work on more than three tickets at once. Before they could start a new ticket, they needed to finish an old one first.*

# MEASLES

*When someone needed help from a veteran when all of their avatars were taken, they had to wait. These tickets were marked with red exclamation marks. The huge number of red exclamation marks on the board made it painfully clear that they had a real problem here: One out of three tickets was blocked! The team said, "Our board has the measles!" Once they saw the problem, they felt pressured to look for a solution. They decided to establish a new policy: The veterans would not be allowed to work alone on any task. Instead, they would work in pairs with newbies until their expertise had spread across the whole team. Justin observed that there was much more collaboration across specializations than ever before!*

# MORE INSIGHTS

*The more Justin explored WIP limits, the more he learned. For example, a colleague from a different business unit told him about their Scrum implementation. During the Sprint Planning, the team decides how many items they will pull from the Product Backlog for the next sprint. Nobody is allowed to push more items into the sprint. Justin' realized that this was actually another kind of WIP limit. Here, the protected, time-boxed sprint works as a limitation mechanism, so it prevents the teams from getting overloaded. Justin thought, "Why wouldn't it be possible to combine this concept with other kinds of WIP limits?" He decided to try this in a future project.*

# TOO MANY PROJECTS

*Things continued to improve for his team, and they were on their way to becoming a Kaizen culture—a culture of continuous improvement. But Justin wanted more. It seemed to him that his organization was working on too many projects at the same time, which was creating the same overload effects he had observed in his team, but at a higher level. How about implementing a similar system, and limiting the WIP on portfolio level? The main difference would be that tickets would represent whole projects instead of features. He talked to some product managers about his ideas and they agreed to give it a try.*

# PORTFOLIO KANBAN

*Of course, their new portfolio board looked different. The columns were called Idea, Vision, Envisioning, Development, and Live. The tickets were color-coded, so that every department had its own color. Now everybody could see at a glance how much capacity the organization was spending on each product. Justin thought that this board would need column limits as well, and he already had some ideas about this. But the product managers didn't want this kind of limitation—for whatever reason. So, there was a need for another kind of limitation. The product managers met once a week and talked about the board and the tickets. Justin facilitated these meetings. Every time someone wanted to start a new project, Justin asked, "Do we have the extra capacity for doing this right now?" And, if not, "Which of the already running projects do we want to stop in order to start a new one?" These questions led to interesting and fruitful discussions. Justin felt that they were making better decisions now and that they had improved collaboration across department borders. So what they actually were implementing was a discussion-based WIP limitation.*

Idea
Vision
Envisioning
Dev.
Live

# WRAP UP

*Why should we limit our Work in Progress?*

# WRAP UP

*How can we limit our Work in Progress?*

# RECOMMENDED READING

***Brain Rules***
*12 Principles for Surviving and Thriving at Work, Home and School*
By John Medina

***Essential Kanban Condensed***
By David J Anderson and Andy Carmichael

***Kanban***
*Successful Evolutionary Change for Your Technology Business*
By David J. Anderson

***Kanban from the Inside***
*Understand the Kanban Method, connect it to what you already know, introduce it with impact*
By Mike Burrows

***Replenishment***
*A Kanban Single*
By Markus Andrezak and Arne Roock (www.kanban-kata.com)

***The Principles of Product Development Flow***
*Second Generation Lean Product Development*
By Donald G. Reinertsen

*www.leankanban.com*
*edu.leankanban.com*
*shop.leankanban.com*
*www.it-agile.de/kanban*

IMPRINT:



*Text: Arne Roock*
*Artwork: Claudia Leschik*
*Layout: Jasna Wittmann/Vicki Rowland*

*First U.S. Edition 2012*
*ISBN: 978-0-9853051-6-1*

*Do you like this booklet?*
*You can order your own copy at*
*shop.leankanban.com*
*www.it-agile.de/justin (Europe)*

leankanban.com

## About Lean Kanban, Inc.

Lean Kanban, Inc. (LKI) is dedicated to developing and promoting the principles and practices of the Kanban Method to achieve the highest quality delivery of professional services through using Kanban. LKI programs include professional development training, a certified training curriculum, events, and published materials.

### Certified Kanban Training

Lean Kanban University (LKU) offers a complete curriculum of certified Kanban classes ranging from introductory to advanced, as well as enterprise services. Visit **edu.leankanban.com** to find certified Kanban training or a knowledgeable and experienced coach or trainer in your area. Additionally, Lean Kanban Services offers private training, coaching, and consulting worldwide.

### Credentialing Programs

LKU provides leadership training for managers, coaches, and trainers. Professional designations include Team Kanban Practitioner, Kanban Management Professional, Kanban Coaching Professional, and Accredited Kanban Trainer. Certified training helps to raise the level of your Kanban expertise and enables you to earn your professional credential.

### Global Conference Series

Join the global Kanban community with Lean Kanban events. The Lean Kanban events series focuses on providing pragmatic, actionable guidance for improving business agility and managing risk with Kanban and related methods. Visit **conf.leankanban.com** for a calendar of upcoming conferences and events.

CPSIA information can be obtained
at www.ICGtesting.com
Printed in the USA
LVRC011427270221
680125LV00017B/166